MAMMALS

Rebecca Woodbury, Ph.D., M.Ed.

Gravitas Publications Inc.

Mammals

Illustrations: Janet Moneymaker

Copyright © 2025 by Rebecca Woodbury, Ph.D., M.Ed.

All rights reserved. No part of this publication may be reproduced, stored in a retrieval system, or transmitted, in any form or by any means, electronic, mechanical, photocopying, recording, or otherwise, without prior written permission from the publisher. No part of this book may be reproduced in any manner whatsoever without written permission.

Mammals
ISBN 978-1-953542-30-4

Published by Gravitas Publications Inc.
Imprint: Real Science-4-Kids
www.gravitaspublications.com
www.realscience4kids.com

Photo credits, AdobeStock: Cover and Title Page: Janet; Above, Juha Saastamoinen; P.3. Rita Kochmarjova; P.5. Happy monkey; P.7. exsodus; P.11. Juha Saastamoinen; P.15. sergojpg; P.16. Aggi Schmid; P.17. Leoniek; P.21. sonya etchison

Do you have a cat or a dog?

Do you live on or near a farm where there are cows or goats?

Maybe you live near a cave and see bats flying out at night.

Cats, dogs, cows, goats, and bats are all **mammals.**

Mice are mammals too.

A **mammal** is an animal that breathes air, feeds milk to its babies, and has hair for at least part of its life. Most mammals give birth to live babies. Only a few lay eggs.

There are many different types of mammals.

Some mammals, like cows and goats, eat only grass and other plants. These mammals are called **herbivores**.

Some mammals, like cats, eat only meat. These mammals are called **carnivores**.

Are there cheese-ivores?

Some mammals, like whales, live only in the ocean.

Some mammals, like squirrels and bobcats, live only on land.

Some mammals, like beavers, spend time both on land and in water.

I avoid swimming.

Did you know that

you are a mammal?

How to say science words

carnivore (KAHR-nuh-vawr)

herbivore (UHR-buh-vawr)

mammal (MAA-muhl)

science (SIY-uhns)

www.ingramcontent.com/pod-product-compliance
Lightning Source LLC
Chambersburg PA
CBHW041632040426
42446CB00022B/3488